OXFORD CONNECTIONS

Changing State

David Glover

Series editor Sue Palmer

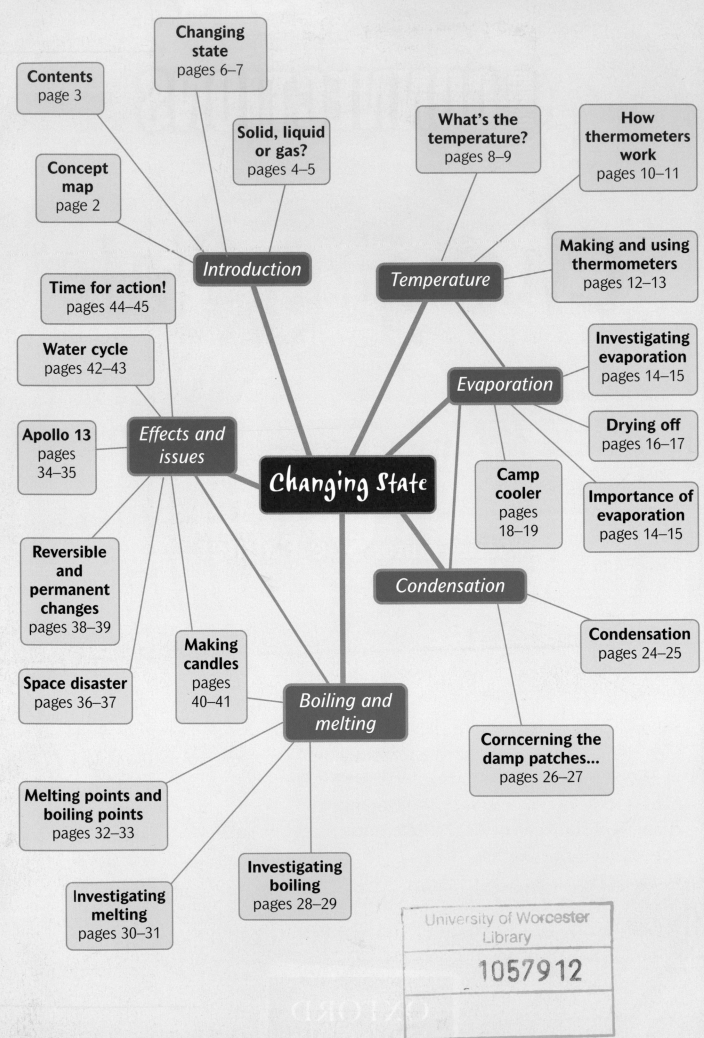

Changing State

Introduction

Temperature

Evaporation

Condensation

Effects and issues

Boiling and melting

Contents

Solids, liquids and gases are all around us and they change from one state to another all the time. Have a look at the story of Apollo 13 in this book – it really shows how understanding changes of state can make the difference between life and death.

Professor Kathy Sykes
Bristol University
Cheltenham Science
Festival Director

Solid, liquid or gas?

Matter

Everything is made from **matter**. Matter is a general word for all the different materials and substances that make up the world. Rock, metal, plastic and glass are matter, and so are soil, water and air. Living things are matter, as are wood, leather and all materials that grew as parts of plants and animals.

All materials are forms of matter.

Rock – a solid

Seawater – a liquid

A **geyser's** steam – a gas

Although there are thousands of different substances, matter usually takes one of just three forms: **solid**, **liquid** or **gas**. These forms of matter are called 'states'. Water flowing in a river is in the liquid state. The state of a stone or an **iceberg** is 'solid'. The air in the atmosphere is a gas.

Each state has its special properties. A solid has shape: it can be picked up and turned around. A liquid flows and changes shape: it takes the shape of the container into which it is poured. A gas can be squeezed into a smaller **volume**: this happens, for example, when a bicycle tyre is inflated with a pump.

We come across three states of matter every day of our lives.

Flour is a solid that can be poured.

Clouds are made of liquid water drops or solid ice crystals that fall to the ground as rain and snow.

Sometimes it can be difficult to decide which state a substance is in. A powder like flour can pour like a liquid. A cloud can look solid, like a ball of cotton wool. However, careful observations show that flour is made from many tiny solid particles, while the cloud is made from tiny drops of liquid water floating in air.

Solid
- Has fixed size and shape

Examples:
metal coins, clay bricks, ice cubes, cool chocolate, wooden chairs

Liquid
- Pours and flows
- Takes the shape of its container

Examples:
tap water, cooking oil, mercury, hot chocolate, blood, paraffin

STATES OF MATTER

Gas
- Mostly invisible. Some coloured, e.g. chlorine is green
- Expands to fill container, e.g. perfume when released in room
- Can be compressed (squeezed) into a smaller volume, e.g. pumping up a tyre

Examples:
oxygen, nitrogen, carbon dioxide, chlorine

CHANGING STATE

Water is a single substance, but it exists in three different states. At normal room temperature, around 20 °C, water is **liquid**. Liquid water can be made to change its state by heating or cooling. These experiments do not need a special laboratory – an ordinary kitchen will do.

Above 100 °C water is a **gas**.

Cooling

🌡 Place liquid water in the freezer.

🌡 When its temperature falls to 0 °C it **freezes** to solid ice.

Heating

🌡 Heat water in a kettle.

🌡 When its temperature rises to 100 °C it **boils**, changing state into a gas called steam.

At 20 °C water is a liquid.

Other substances change state too, but at different temperatures. Chocolate, for example, **melts** at about 36 °C; cooking oil boils at about 300 °C.

Chocolate changes state from solid to liquid at about body temperature. It melts in your mouth!

Below 0 °C water is a **solid**.

The oil in a fish-fryer is much hotter than boiling water.

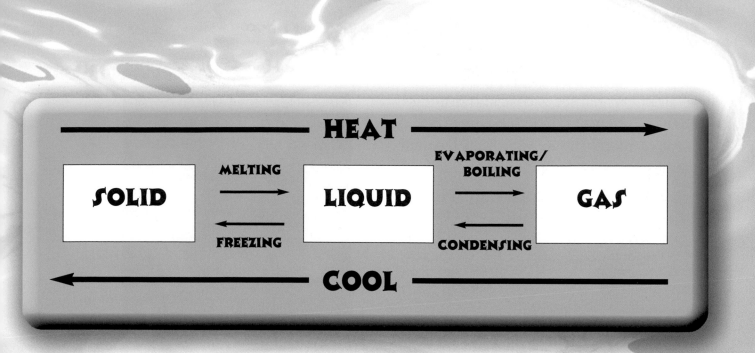

HEAT →

SOLID → **MELTING** → LIQUID → **EVAPORATING/ BOILING** → GAS

← **FREEZING** ← **CONDENSING** ←

← **COOL**

The diagram above shows the changes of state that can happen to water and other substances. Changes of state have different names, depending on the start and finish states. Melting, for example, is the change of state from solid to liquid.

heat from Sun

evaporation

water puddle

Puddles disappear because of evaporation.

evaporation

A liquid can change to a gas both by **evaporation** and by boiling. Water evaporates into the air at all temperatures. At room temperature this change is quite slow, but evaporation speeds up when the water is heated. Eventually the change becomes so fast that steam bubbles grow inside the liquid. The water is then boiling.

Boiling water changes state quickly and will soon boil dry.

What's the temperature?

We all know that ice is cold and boiling water is hot, but how do we know their temperatures? Who decided that ice **melts** at 0 °C and water **boils** at 100 °C?

Instruments that measure temperature are called thermometers. The thermometer is usually said to have been the invention of the great Italian scientist Galileo Galilei (1564–1642). Galileo is best known for his discoveries about **gravity**, and for being the first scientist to use a telescope in **astronomy**.

Galileo made his first thermometer in 1597. The diagrams on page 10 show how it worked. A glass bulb is connected to long tube. The bulb and tube are turned upside down and the bulb is warmed in the hands. The mouth of the tube is held under the surface of some water in a jar. The hands are removed. As the air inside the bulb cools, it contracts and water rises further up the tube. Temperature changes in the surroundings then make the water level in the tube rise or fall.

The changing water level in Galileo's thermometer showed how the temperature was changing, up or down. But it did not give the temperature as a number that could be quoted as 'the temperature'. Scientists soon realized that, just as a ruler measures lengths on a scale of units (centimetres for example), a thermometer scale needed units for temperature measurement.

Galileo Galilei

A famous story (for which there is no historical evidence) tells how Galileo dropped cannon balls from the leaning tower in Pisa. He observed that all objects fall at the same rate, whatever their size.

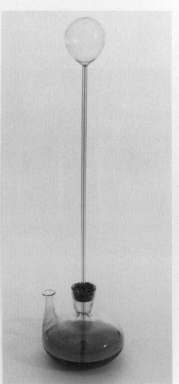

You can find this copy of Galileo's thermoscope in the Museum of the History of Science in Florence, Italy.

One of the first useful temperature scales was invented by the German scientist Daniel Gabriel Fahrenheit (1686–1736). Fahrenheit used a mercury thermometer (see page 11) to make his scale.

Daniel Gabriel Fahrenheit

He realized that he needed to mark two fixed temperatures on his thermometer that could be reproduced by scientists in other places. The lower temperature was the coldest thing he knew of – an equal mixture of salt and ice. This he labelled 0 °F. The upper temperature was something consistently warm – the human body. This he labelled as 96 °F.

The scale was then made by dividing the length of the thermometer tube between these two 'fixed points' into 96 equal degrees. (Later Fahrenheit adjusted his scale slightly. He fixed the melting temperature of **pure ice** at 32 °F and the boiling point of water as 212 °F. On the new scale there are exactly 180 degrees between the two fixed points. The temperature of the human body on the adjusted scale is 98.4 °F.)

Fahrenheit's scale proved very successful and is still often used. However his choice of 32 °F and 212 °F for the melting and boiling temperatures of water was more complicated than is necessary. In 1742 the Swedish scientist Anders Celsius (1701–1744) took the obvious next step. On his thermometers he set the temperature of melting ice as 0 °C and the temperature of boiling water as 100 °C. Because the scale was divided into a 100 equal degrees it was called a 'centigrade' scale. Each interval was equal to 1 °C (one degree centigrade). The scale is still in use today and we talk about 'degrees Celsius' to honour its inventor.

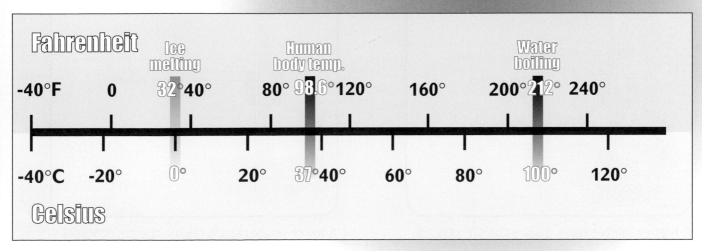

Fahrenheit and Celsius scales compared

How thermometers work

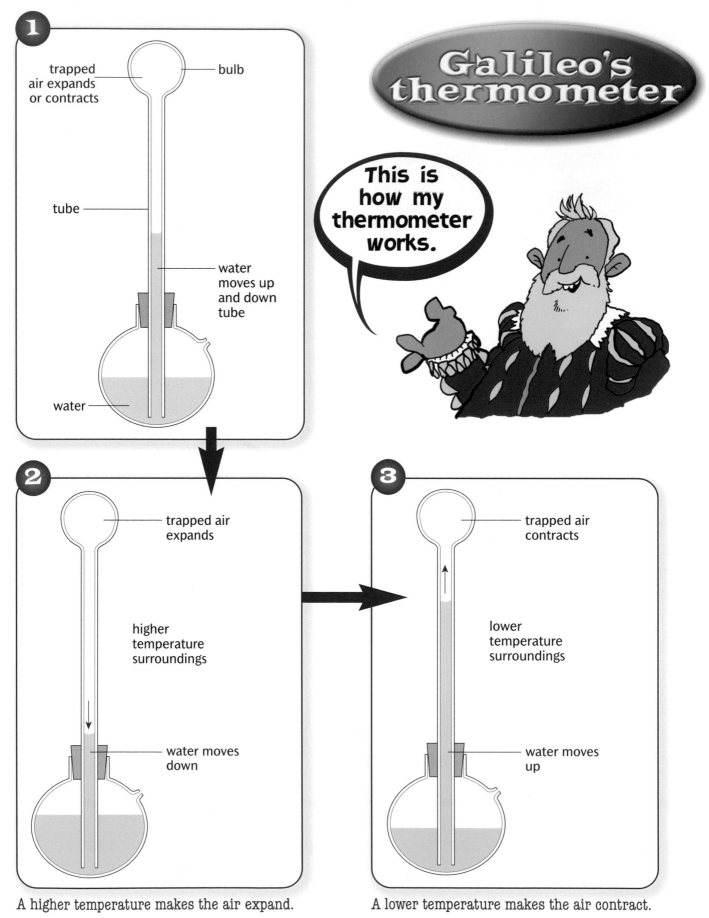

1

trapped air expands or contracts

bulb

tube

water moves up and down tube

water

Galileo's thermometer

This is how my thermometer works.

2

trapped air expands

higher temperature surroundings

water moves down

A higher temperature makes the air expand.

3

trapped air contracts

lower temperature surroundings

water moves up

A lower temperature makes the air contract.

A mercury thermometer

This is how a mercury thermometer works.

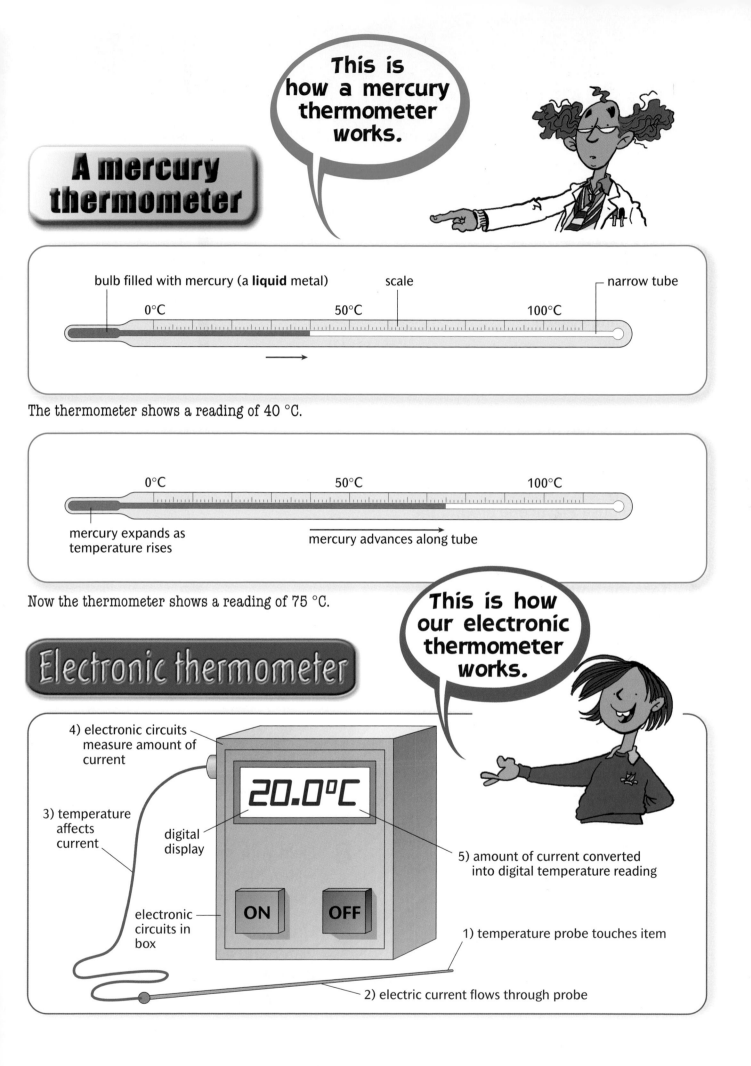

bulb filled with mercury (a **liquid** metal) scale narrow tube

0°C 50°C 100°C

The thermometer shows a reading of 40 °C.

0°C 50°C 100°C

mercury expands as temperature rises

mercury advances along tube

Now the thermometer shows a reading of 75 °C.

Electronic thermometer

This is how our electronic thermometer works.

4) electronic circuits measure amount of current

20.0°C

3) temperature affects current

digital display

5) amount of current converted into digital temperature reading

electronic circuits in box

ON OFF

1) temperature probe touches item

2) electric current flows through probe

How to make an air thermometer

You will need:

Narrow plastic tube

Plasticene

Strip of card, marked with a scale

500 ml plastic lemonade bottle

Coloured water in a plastic beaker

1 Push one end of the plastic tube 2 or 3 cm down inside the neck of the plastic lemonade bottle.

2 Firmly push Plasticene around the tube at the neck of the bottle to seal the tube in position.

3 Turn the bottle upside down. Squeeze the bottle gently, then dip the free end of the tube in the coloured water.

4 Partly release the bottle to draw a drop of coloured water up into the tube.

5 Pull the tube from the water, then release the bottle fully so that the drop moves to the middle of the tube (you may need to try this procedure several times before you get the drop in the right position).

6 Turn the bottle and tube back upright.

7 Draw a scale on a card strip and tape the card strip to the tube.

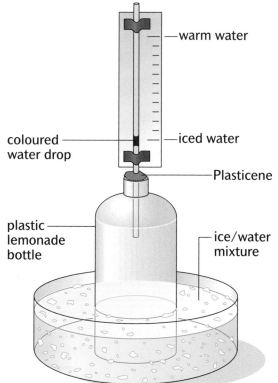

warm water

coloured water drop

iced water

Plasticene

plastic lemonade bottle

ice/water mixture

8 Hold your thermometer in warm water. Mark the position of the coloured water drop on the card.

9 Hold your thermometer in iced water and again mark the position of the coloured water drop. Observe how the drop has moved.

1. Stir
stirring stick
water

2. Immerse thermometer
surface of water
thermometer bulb

3. Wait until the liquid in the tube has stopped rising.

4. Take reading

Do not use the thermometer as a stirrer, it may break. Use a spoon or a plastic rod.

Measuring water temperature

It is important to follow the correct procedure when using a thermometer. If you do not, your measurement will not be accurate. For example, to measure water temperature with a liquid-in-glass thermometer you should proceed as above.

thermometer

✗ Immerse the thermometer fully. ✓

✗ Do not remove the thermometer to take the reading. ✓

Digital thermometer

This is the procedure for measuring water temperature with a digital thermometer.

1. Stir

2. Immerse
38.0°C
ON OFF

3. Wait

4. Read
steady reading
45.5°C
ON OFF

5. Record

	Results
1	45.5°C
2	

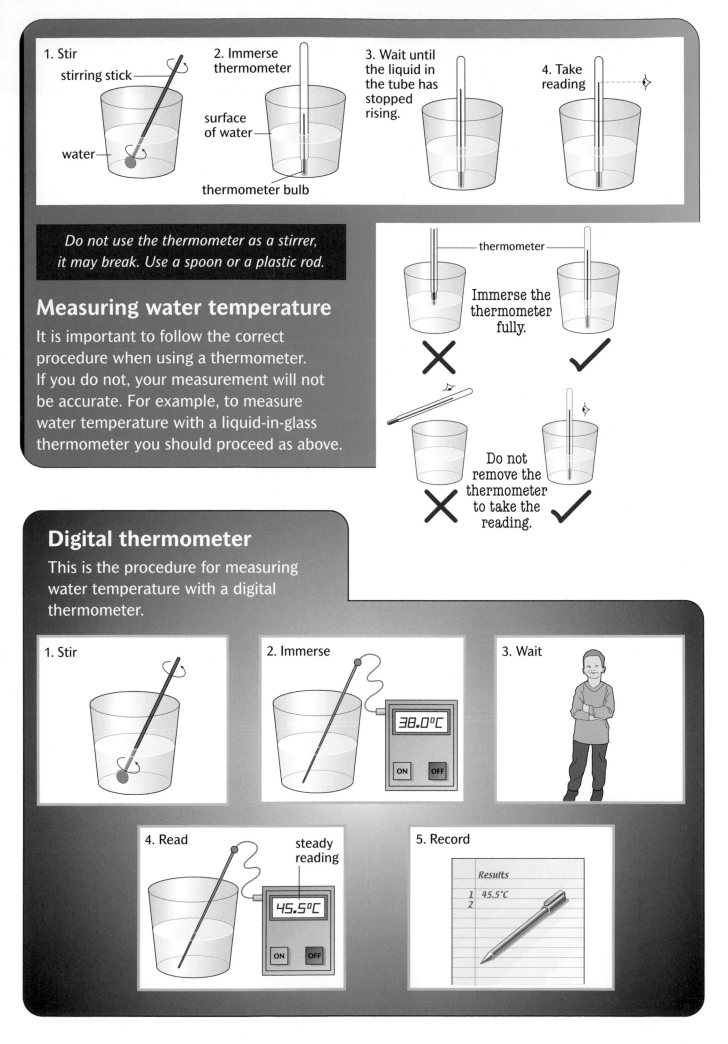

Investigating evaporation

Case Study

Introduction

Evaporation is the change of state when a liquid turns into a gas. Evaporation explains why wet clothes dry out. The water **evaporates** into the air. In this investigation we wanted to find out how the speed of evaporation is affected by:

a) the temperature and b) a breeze

We predicted that evaporation would be faster at a higher temperature. We also predicted that evaporation would be faster when there is a breeze than when the air is still.

Equipment and method

The equipment we used is listed below.

- Plastic trays
- Thick card
- Stopwatch
- Digital thermometer
- Electric fan
- Bucket of water
- Scissors

To test the effect of temperature we compared the drying times for three card squares that were first soaked in water then kept at different temperatures. To make our test fair we cut each card square the same size. We dipped each card in the water bucket for 10 seconds so that it soaked up the same amount of water. We laid the wet cards on plastic trays.

We put the first tray in the bottom of a fridge, the second tray in a cool outdoor shed and the third tray in a warm room. We made sure that there was no breeze in any of the places, so that the test was fair. Then we measured and recorded the temperature at each place with the digital thermometer. After that, we checked the cards in the trays every 10 minutes to see if they were dry. Table 1 gives our results.

To check how a breeze affects evaporation we used a similar method. This time we set up two wet card samples in the classroom so that they were at the same temperature. We put the first card in a place where the air was still. To create a 'breeze' over the second card, we blew air over it using a fan. As before, we checked every 10 minutes to see if the cards were dry. Table 2 gives the results of this experiment.

Results

Table 1 The effect of temperature on evaporation

card	location	temperature	drying time
1	fridge	2°C	still wet after 6 hours
2	shed	11°C	3 hours 50 minutes
3	classroom	22°C	40 minutes

Table 2 The effect of a breeze on evaporation

card	conditions	temperature	drying time
4	still air	22°C	40 minutes
5	breeze from fan	22°C	20 minutes

Conclusions

The results in Table 1 confirm our prediction that evaporation is faster at a higher temperature. The card in the warm classroom dried out first. The card in a cold fridge did not dry out at all during the day.

The results in Table 2 confirm our prediction that a breeze speeds up evaporation. The card near the fan dried out twice as fast as the card in still air.

Drying Off

A good drying day is warm and windy. This is because the warmth makes the water evaporate quickly, and the wind carries the **water vapour** away. To dry well the clothes must be spread out as much as possible. Spreading the clothes out well is better than leaving them folded because it maximizes the surface area from which the water can escape. Folded sheets and buttoned shirts do not dry as well as sheets and shirts that are pegged out properly.

Sadly, some days are not so good for drying the washing. Even if it is not raining it may be just cold, still and damp. Then washing stays damp too. On days like these a tumble drier does the trick. The diagram on page 17 shows how it works.

Moving air dries clothes quickly.

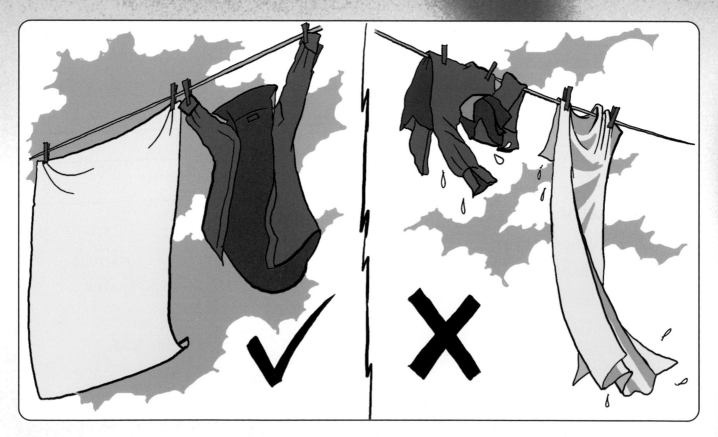

How to make sure your washing dries quickly

HOW A TUMBLE DRYER WORKS

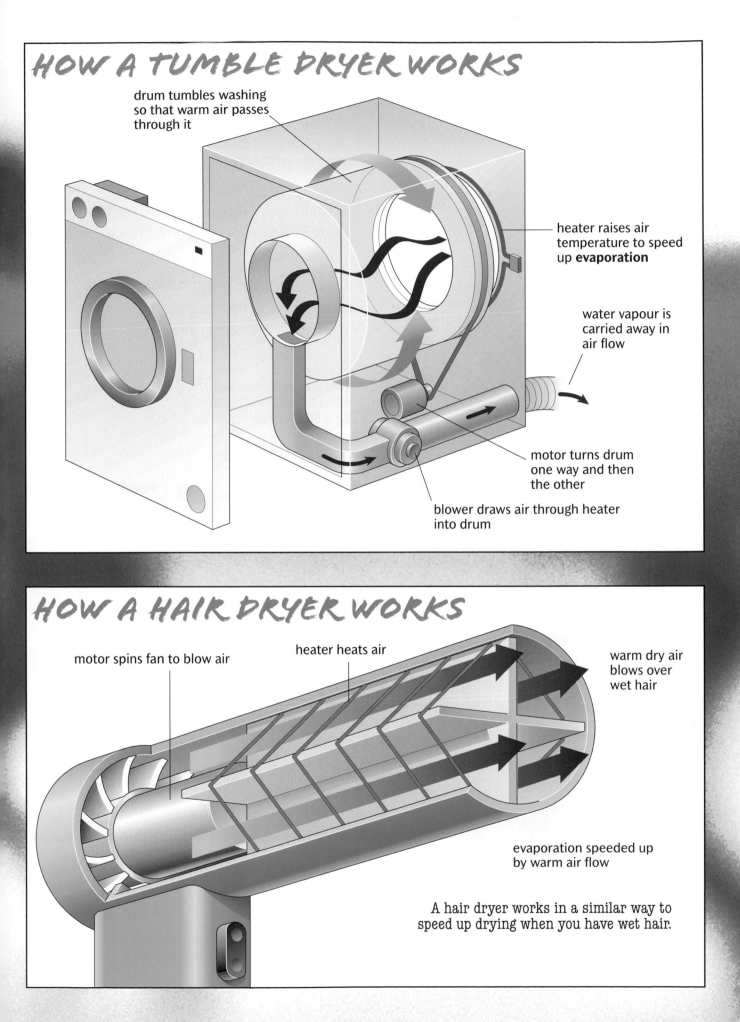

drum tumbles washing so that warm air passes through it

heater raises air temperature to speed up **evaporation**

water vapour is carried away in air flow

motor turns drum one way and then the other

blower draws air through heater into drum

HOW A HAIR DRYER WORKS

motor spins fan to blow air

heater heats air

warm dry air blows over wet hair

evaporation speeded up by warm air flow

A hair dryer works in a similar way to speed up drying when you have wet hair.

CAMP COOLER

When a **liquid evaporates** it cools. It carries away heat. This is why you feel cold and shivery when you get out of the sea or a swimming pool in a breeze. The evaporating water cools your skin. It actually feels warmer in the sea because there is no evaporation from your skin under water.

Our bodies use evaporation to keep cool when we are working hard. Sweat carries heat away by evaporating.

Getting out of the sea after a warm swim can be a chilly experience.

As water evaporates from an elephant's skin, it takes some of the elephant's body heat with it.

Elephants know about the cooling effect of evaporation too. They don't just lie in the water when they are hot. They spray themselves with water from their trunks to speed up evaporation and increase the cooling effect.

You can use evaporation to make a drinks cooler when you are camping. These diagrams give two designs.

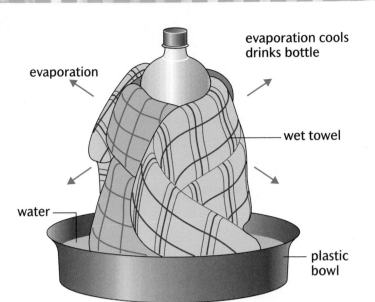

evaporation

evaporation cools drinks bottle

wet towel

water

plastic bowl

Wet towel cooler

✪ Evaporation from towel cools drinks cans and bottles.

✪ Place in the shade where there is a strong breeze.

✪ Make sure the towel does not dry out.

Porous pot cooler

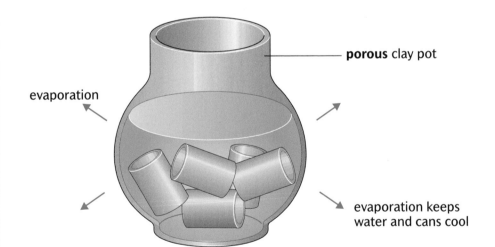

porous clay pot

evaporation

evaporation keeps water and cans cool

This type of cooler was used in ancient times.

✪ Evaporation keeps water and drinks cans in pot cool.

Can you make and test an evaporation cooler?
How could you improve the design?
How could you make sure the cooler stays in a steady breeze to increase evaporation?

The importance of evaporation

It's not just water that **evaporates**. Other **liquids** do too. Petrol, for example, evaporates more quickly than water. The smell of spilt petrol at a garage spreads rapidly through the air. Petrol vapour is very dangerous as a glowing cigarette or an electric spark can make it explode.

Evaporation is important in many ways. This diagram **summarizes** important inventions and processes linked to evaporation. Evaporation hazards are described in more detail on pages 22 and 23.

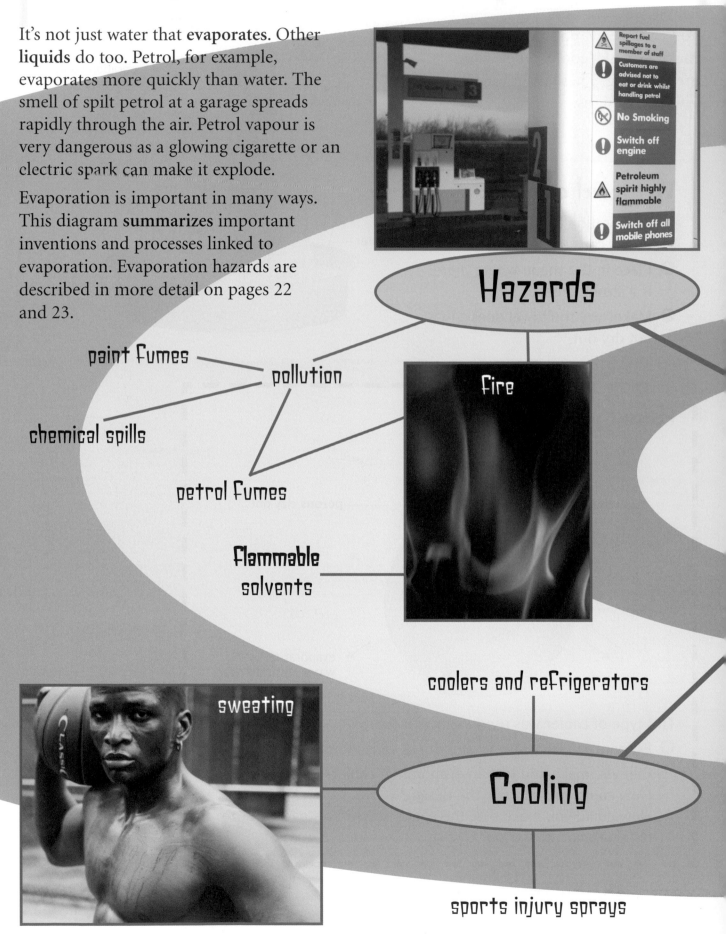

Report fuel spillages to a member of staff

Customers are advised not to eat or drink whilst handling petrol

No Smoking

Switch off engine

Petroleum spirit highly flammable

Switch off all mobile phones

Hazards

paint fumes

pollution

chemical spills

petrol fumes

flammable solvents

Fire

coolers and refrigerators

sweating

Cooling

sports injury sprays

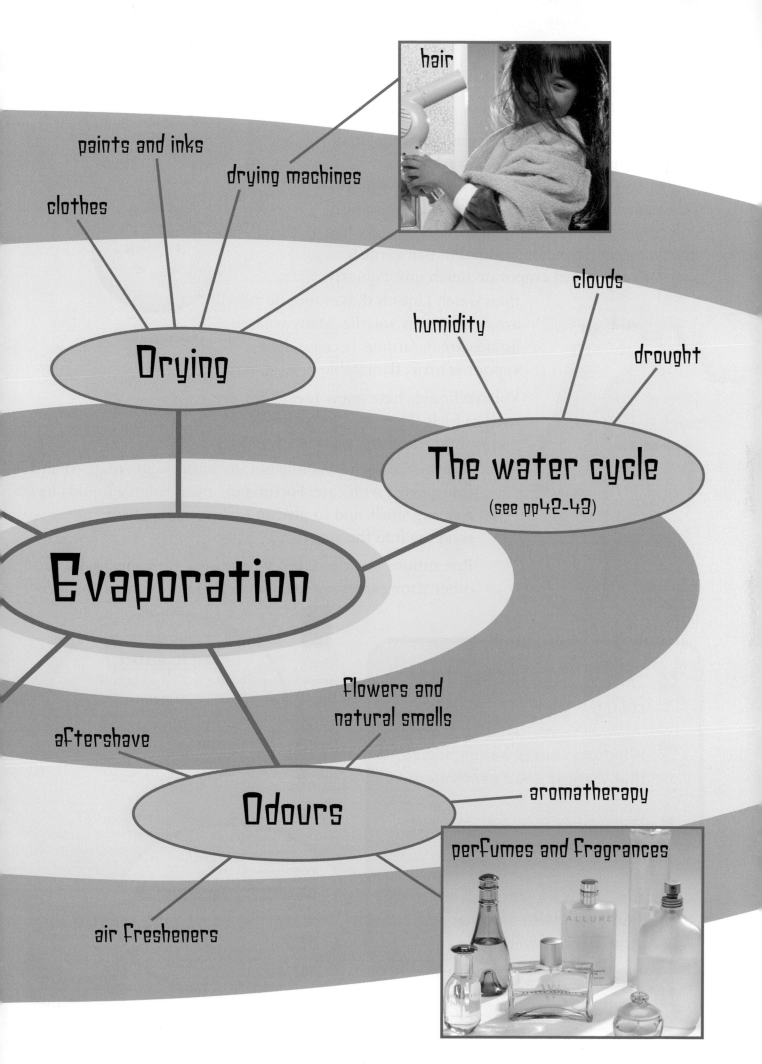

hair

paints and inks

drying machines

clothes

Drying

clouds

humidity

drought

The water cycle
(see pp42-43)

Evaporation

flowers and
natural smells

aftershave

aromatherapy

Odours

perfumes and fragrances

air fresheners

Evaporation hazards

All **liquids evaporate**. They change from liquid to gas and escape into the air. At everyday temperatures, water evaporates slowly. We cannot see or smell **water vapour** and it creates no hazard. Other liquids are different. Liquids such as alcohol, methylated spirits, acetone (used in nail varnish remover) and petrol evaporate much more quickly than water. Liquids that evaporate rapidly are described as **volatile**. Many volatile liquids are hazardous because their vapour is toxic, **flammable** or explosive.

Volatile liquids have many uses. They are used as fuels. They are used to make paints, inks, varnishes and cleaning fluids.

To avoid accidents volatile liquids must be stored, used and transported with care. Fortunately most volatile liquids have a strong smell, and so give an early warning if they are escaping into the air.

Precautions must be taken when using varnishes, paints and other strong smelling substances at school and in the home:

☠ **They must not be used in a confined space.**

☣ **Windows should be opened so that there is good ventilation.**

🧯 **The vapour must be kept well away from naked flames.**

🔒 **They must be stored in approved containers with lids that seal properly.**

quickly and effective

CAUTION: HIGHLY FLAMMABLE

Keep away from heat and flame. Avoid prolonged contact with the skin and wash hands after use. Avoid contact with polished surfaces and synthetic materials. Keep out of reach of children. roduct contains Bit-

SAINSBURY'S NAIL POLISH REMOVER With Conditioners 250 ml ℮

t covering power and se

FLAMMABLE
Keep out of the reach of children.
Use only in well ventilated areas.

SAFETY, HEALTH AND ENVIRONMENT

SAFETY
Keep away from sources of ignition - no smoking.
When empty do not use this pack to store foodstuffs.

HEALTH
Ensure maximum ventilation during application and dry by opening available windows and doors. estrict interior use to small surface areas such a ng boards and window frames. mended for interior use on

BERGER
PURE BRILLIANT WHITE
Undercoat
the perfect foundation for liquid gloss

Hazard warnings must be displayed on bottles and cans that contain volatile liquids.

There are strict regulations for handling volatile fuels. For example:

- Petrol pumps must only be operated by persons over 16 years of age.

- Fuel must only be carried in approved containers.

- Smoking, matches and sources of electrical sparks are prohibited in locations where fuel is dispensed or stored.

- Spills must always be reported and cleared up correctly.

Many serious accidents have been caused by the evaporation of volatile liquids in confined spaces. For instance factory workers have been overcome by fumes in storage tanks, and petrol vapour from leaking fuel pipes has caused explosions on pleasure boats. These accidents can be avoided if people are aware of the dangers and follow the regulations.

Leaking petrol vapour caused an explosion aboard this pleasure boat.

Firefighters wear breathing apparatus and protection suits to clean up a chemical spill.

Condensation

As the night air cools, water vapour condenses. Dewdrops form on leaves.

Condensation is the reverse of evaporation. It is the change of state when a **gas** is cooled and becomes a **liquid**.

The atmosphere holds **water vapour evaporated** from rivers, lakes, wet soil and other water sources. Warm air can hold more water vapour than cold air. This explains why, when warm air touches a cold surface, drops of liquid water **condense** on the surface. For example, condensation occurs when you run a hot shower in a cold bathroom. Water evaporating from the shower condenses on nearby cold surfaces. This means that windows and mirrors 'steam up' with condensation.

warm moist air + cold surface ➔ condensation

Condensation on cold windows

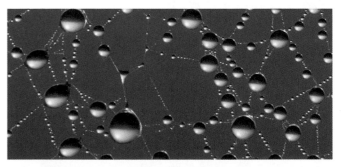

Dew drops on a spider's web

Morning mist (condensation in the air over cold ground)

Condensation on a cold can in a warm room

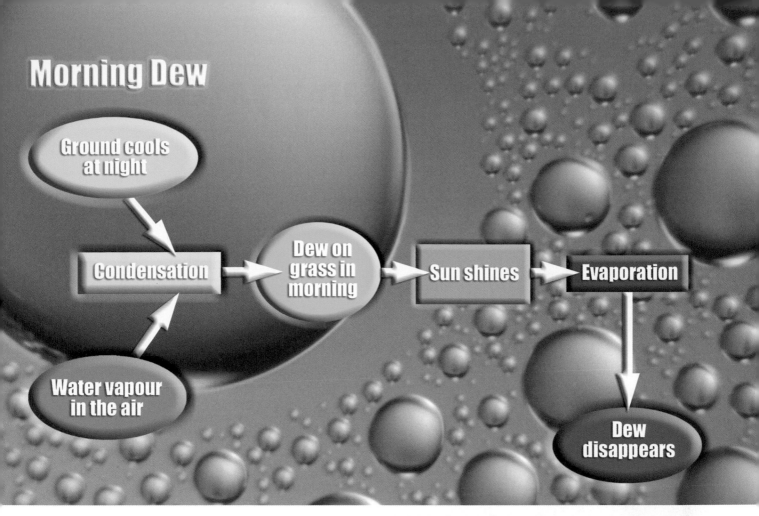

Morning Dew

Ground cools at night → Condensation

Water vapour in the air → Condensation

Condensation → Dew on grass in morning → Sun shines → Evaporation → Dew disappears

Dew ponds

In medieval times farmers dug saucer-shaped ponds to collect the dew. They lined the ponds with straw, then covered the straw with clay and pebbles. As the bottom of the dew pond was shaded from the rising Sun, the dew did not evaporate so quickly. The clay stopped dew and rainwater leaking away. These dew ponds were a water source for their animals.

A dew pond on the South Downs

At night, water condenses on the cold pebbles and collects at the bottom of the dew pond.

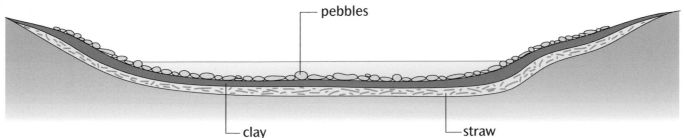

pebbles

clay

straw

Cross-section of a dew pond.

Concerning the damp patches...

Mr M Adams
Brand New Homes Ltd
Bricklayers Yard
The Trading Estate
Dry Drayton
Cambs
CA4 7ET
01/04/03

30 Watery Way
The Deepings
Pondlesbury
PT2 8TZ

Dear Mr Adams,

Following our recent telephone conversations I felt I must put my concerns about the damp problems in our new house in writing.

As you know we moved in four months ago, and at first we were very pleased with the house. However, as various problems have appeared (which I have listed in my previous letters) we have become more and more dissatisfied with the quality of the workmanship. Now there are damp patches on the walls in the kitchen and bathroom, and mould growing around the window frames.

I understand that the building inspector will be visiting in the next few days to look at the damp and other problems. We look forward to receiving his report. I must say that as things stand we are extremely unhappy with the situation. We feel strongly that your company should be both obliged to correct all the faulty work immediately, and to offer us substantial compensation for all the inconvenience caused.

I look forward to your prompt reply.

Yours sincerely

A Swan

BRAND NEW HOMES LTD

Mr A Swan
30 Watery Way
The Deepings
Pondlesbury
PT2 8TZ
7 April 2003

Bricklayers Yard
The Trading Estate
Dry Drayton
Cambs
CA4 7ET

Dear Mr Swan,

Thank you for you letter dated 1/4/2003 concerning the damp patches in your new kitchen and bathroom. We have now received Mr Smith's inspection report. As you will be aware Mr Smith is an independent inspector who has no connection with this company.

Mr Smith states that he can find no evidence for leaks from the plumbing. The windows have been fitted correctly and there is no water entering the building through the walls or roof.

Mr Smith suggests that the damp and mould are due to condensation caused by the recent cold weather and lack of ventilation. To reduce the problem in future he recommends that you take the following steps.

- Keep the kitchen and bathroom properly heated. Condensation is always greater on cold walls and surfaces.

- Make sure the extractor fans are on when you run a bath or cook. They come on automatically when the room light is switched on. The fans help remove the damp air from the room.

- Open a window to improve ventilation when you are cooking or showering.

- Do not let too much steam escape into the air when you are **boiling** vegetables. Cover boiling saucepans with lids, and turn down the hob.

- Dry off any damp surfaces such as tiles or windows with a sponge or cloth when you have finished. This will prevent mould growth.

We hope you find this advice helpful and feel sure that if you take the steps listed the problem will be solved. As you can see Mr Smith has found no faults in the design or construction of the house and has not identified poor workmanship as a problem. I hope you will agree that in these circumstances there is no repair work for us to undertake and no case for compensation.

Please do not hesitate to contact us again if we can be of further help.

Yours sincerely

M Adams

Site Manager

Investigating boiling

Case Study

Investigation notes

Objectives

To use a computer **data-logger** to record the temperature of water as it is heated to boiling point. To investigate how the temperature of water changes as it **boils** and changes state from **liquid** to **gas**.

Apparatus

Computer

Data-logging software

Temperature probe and interface

Saucepan

Stove

Diagram

Safety

Hazard from boiling water and hot stove. Teacher operated the stove and handled the pan.

Method

Pan half filled with cold tap water and placed on stove. Temperature probe submerged in water.

Data-logging software set to record a graph of temperature over a 15 minute period and started.

Stove switched on.

Teacher stirred water continuously as it heated up.

Results

Results are shown in graph below

Water boiling

(graph: temperature degrees Celsius vs time in minutes; curve rises from about 15°C at 0 minutes, steadily increasing to about 99°C around 6–7 minutes, then levelling off to 12 minutes)

Observations and conclusions

- The water started at 20 °C
- It heated up steadily to 99 °C then began to boil
- The water temperature did not rise above 99 °C as it boiled
- The boiling point of pure water is 100 °C. Our sample boiled at a lower temperature. Either our thermometer is not accurate, the water was not pure or something else affected the temperature (our teacher said that the air pressure might have made a difference).

Write up

We plan to write up these notes as a report using the following structure.

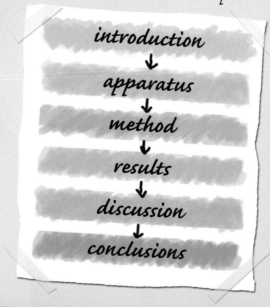

introduction
↓
apparatus
↓
method
↓
results
↓
discussion
↓
conclusions

Investigating melting

Melting is the change of state when a **solid** changes to **liquid**. Experiments show that when liquid water **boils** and changes to the **gas** steam, its temperature does not change. It boils at a steady temperature of about 100 °C. Does melting happen the same way? Does the temperature of the ice stay the same as it melts, or does it change? What is the melting temperature?

Your task is to plan an investigation of melting ice. Your investigation should show how the temperature of ice taken from a freezer changes as it warms up and melts. You have the following apparatus and materials available.

How does the temperature change as the ice melts?

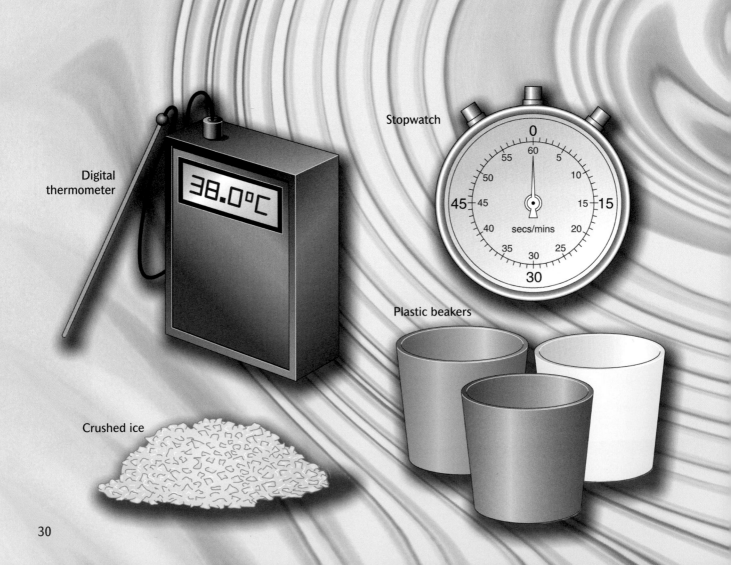

Digital thermometer

Stopwatch

Plastic beakers

Crushed ice

You should write up your plan in the following order.

Introduction

↓

Planned experiment

↓

Apparatus

↓

Recording results

These notes will help you to develop your plan.

Introduction

This should describe the objectives of your experiment. What are you planning to investigate? Why are you making the investigation? Have you any predictions for the outcome of your experiments?

Planned experiment

This should describe the experiment you will make. What are you going to measure? What kind of equipment will you use to make your measurements? How will you set about making the measurements?

Apparatus

This should list all the apparatus and materials you need to make your investigation.

Recording results

This should describe how you plan to record your results. Will you take readings from a thermometer at regular intervals, or will you use a computer **data-logger**? How will the results be displayed – in a table or as a graph? How would you expect your results to appear if your predictions are correct?

A well-planned investigation should produce reliable results.

Melting points and boiling poir

The **melting** point is the temperature at which a **solid** changes to a **liquid**. The **boiling** point is the temperature at which a liquid bubbles and changes rapidly into a **gas**.

Different substances **melt** and **boil** at different temperatures. Water for example melts at 0 °C and boils at 100 °C; the metal iron melts at 1540 °C and boils at 2760 °C.

If we know the melting and boiling points we can work out if a substance is solid, liquid or gas at everyday temperatures. If the melting point is high, as in the case of iron, then the substance is solid. If the melting point is below normal room temperatures, but the boiling point is higher, then the substance is liquid, like water. If the boiling point is lower than normal temperatures, then the substance is a gas; oxygen for example boils at −183 °C.

Molten iron can be poured and shaped before it cools.

The gases that make up the air all have boiling points well below 0 °C.

They can be liquefied (made liquid) by cooling below their boiling temperature.

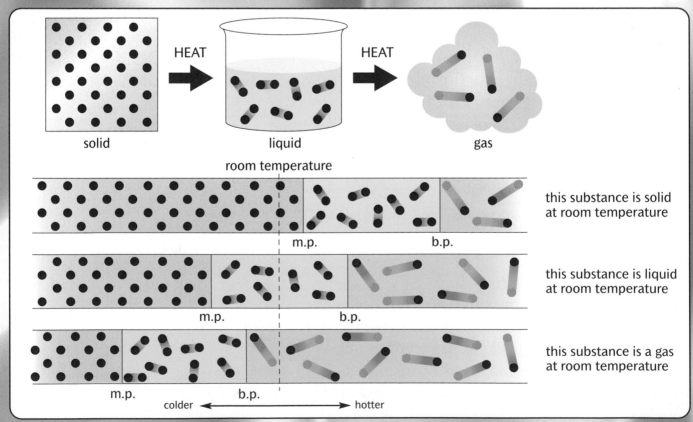

The state of a substance depends on its melting and boiling points in relation to room temperature.

32

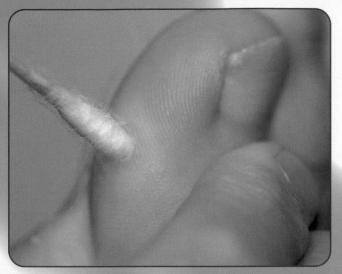

Liquid nitrogen can be used to remove warts.

Nitrogen is a liquid at temperatures below −196 °C.

Liquefied gases have many important uses. The huge tank on a Space Shuttle carries liquid oxygen and liquid hydrogen, which together are the fuel for the main rocket engines. Liquid nitrogen is used whenever very low temperatures are needed. In hospital for example, warts and verrucas are treated by **freezing** them with liquid nitrogen.

Metals have high melting points and are normally solids. But one metal, mercury, is a liquid at room temperature. Mercury's melting point is −39 °C. This is why mercury is suitable for use in thermometers.

Most solids melt if they are made hot enough, even rock. The **lava** that pours from a **volcano** is liquid rock that melted at about 700 °C.

This table lists the melting and boiling points of some substances. The substance with the lowest known melting point is helium (−270 °C). The metal with the highest known melting point is tungsten (3387 °C).

The filament in this bulb is a coil of tungsten wire. It glows white hot.

Table of melting points and boiling points

Substance	Melting point in °C	Boiling point in °C	Normal state at room temperature
water	0	100	liquid
oxygen	−219	−183	gas
helium	−270	−269	gas
nitrogen	−210	−196	gas
iron	1540	2760	solid
tungsten	3387	5420	solid
alcohol	−117	78	liquid
gold	1064	2850	solid
mercury	−39	357	liquid

APOLLO 13

"Successful Failure"

The story of the Apollo 13 has been told in an exciting Hollywood film starring Tom Hanks as mission commander Jim Lovell. But the real life events that took place during six days in April 1970 gripped people around the world more than any action movie. Officially classified as a "successful failure" Apollo 13 was a mission to the Moon that came within a hair's breadth of disaster. Remarkably, for such a complex project, the simple changes of state, melting, boiling and evaporation, played an important role in the unfolding drama.

On 21 July 1969, Neil Armstrong, commander of Apollo 11, became the first human being to set foot on the Moon. Apollo 11 and its follow-up mission Apollo 12, both landed astronauts on the Moon, then returned them safely to Earth.

By the time Apollo 13 was launched on 11 April 1970 the public were beginning to take the success of Moon missions for granted. But, at 9:08 pm on 13 April, some 200,000 miles from Earth, the crew felt a sharp bang followed by a series of vibrations. Lunar module pilot Jack Swigert saw a warning light flash on and calmly reported over the radio link to Earth, "Houston, we've had a problem here".

Looking out of the window of the command module Jim Lovell saw debris and gas evaporating into space. The instruments showed that one of the main oxygen tanks was empty and the second was losing pressure rapidly. An explosion had destroyed one tank and badly damaged the other.

Apollo 13: its launch and ground crew in Houston

The situation was critical – but not yet fatal. The Moon landing was immediately abandoned as the astronauts and ground crew worked frantically to devise a plan to return the crippled craft to Earth. The craft had not been designed with enough power to reverse direction in mid-flight, so it had to continue on its outward journey, swinging around the Moon before heading back to Earth. In the meantime the crew moved into the lunar-landing module to use the oxygen, water and power supplies that it carried.

The crew spent the following days in freezing temperatures with just a cupful of water to drink each day as they conserved their remaining power and water supplies. But the plan went well. A few hours before landing they returned to the dark command module, switched on its electric circuits, and prayed that there would be enough power left for a safe touchdown. With baited breath the ground crew waited. Then, at 12:07 pm on 17 April, nearly four days after the explosion that threatened to maroon them in space, cheering erupted as the crew of Apollo 13 parachuted safely into the Pacific Ocean.

Months later the official report on the mission showed that the following sequence of events had led to the explosion.

Apollo 13 astronauts

Day before launch

- The **liquid** oxygen tank was not working properly.
- Engineers used an electric heater inside the tank to **boil** off liquid oxygen.
- The insulation on a faulty switch melted.
- The problem was not noticed.

Two days after launch

- The astronauts were told to turn on the fan inside the tank to stir the liquid oxygen.
- The faulty insulation caught fire.
- The tank exploded.
- Oxygen from the tank **evaporated** into space.

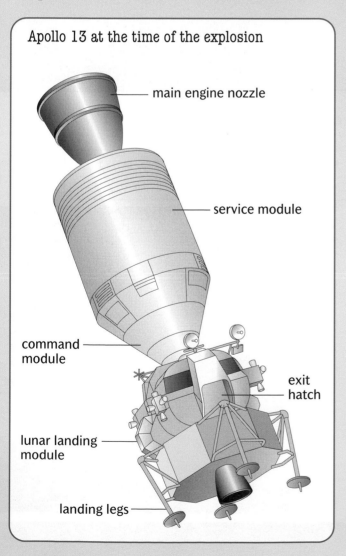

Apollo 13 at the time of the explosion

main engine nozzle

service module

command module

exit hatch

lunar landing module

landing legs

In terms of its goal to land men on the Moon, Apollo 13 was a failure. But because the astronauts and ground crew were able to use their knowledge and skills to overcome disaster without loss of life, it can be safely described as a "successful failure".

SPACE DISASTER

The *Challenger* exploded, killing all its crew.

Space accidents are rare, though there have been some near misses such as Apollo 13, and one terrible disaster. *

On 28 January 1986 the Space Shuttle *Challenger* exploded 73 seconds after lift-off, instantly killing all seven astronauts on board. Around the world people were horrified as they witnessed the explosion on TV. The disaster investigation discovered that a change in state, caused by cold weather at the launch site, had triggered the explosion.

The sequence of events that led to the disaster can be **summarized** as follows.

➡ **January 22–27:**
launch postponed four times due to technical problems/ bad weather

➡ **January 27:**
weather forecast predicted **freezing** temperatures. Engineers wanted to postpone launch again, but managers decided to go ahead next day

➡ **Night-time temperature** dropped to −8 °C. Rubber seals in booster rockets froze

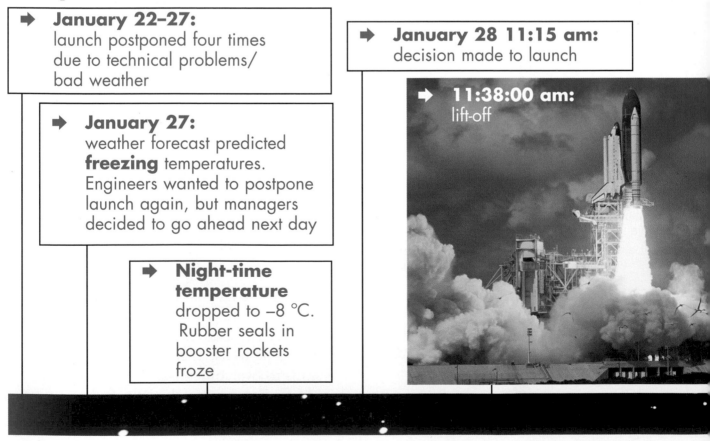

➡ **January 28 11:15 am:** decision made to launch

➡ **11:38:00 am:** lift-off

* Shortly after this text was written a second shuttle, *Columbia*, was lost as it returned to Earth at the end of its mission. At the time of writing, the cause of the disaster was still under investigation.

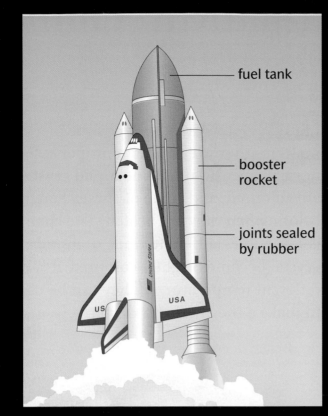

Space shuttle launch

- fuel tank
- booster rocket
- joints sealed by rubber

In a famous demonstration at the disaster enquiry, Nobel prize winning scientist Professor Richard Feynman cooled a piece of rubber seal from the Shuttle's booster rocket in a cup of iced water. The rubber changed state from a springy material to a stiff material. The rubber had frozen.

Professor Feynman

Challenger disaster enquiry

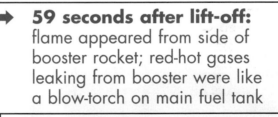

➡ **59 seconds after lift-off:** flame appeared from side of booster rocket; red-hot gases leaking from booster were like a blow-torch on main fuel tank

➡ **1 second after lift-off:** puffs of smoke seen from leaking booster rocket joint

➡ **73 seconds after lift-off:** Space Shuttle *Challenger* exploded

Reversible and permanent changes

Reversible

Changes of state can be reversed. If you take some ice from the fridge it **melts** to **liquid** water. If you put the water back in the fridge it **freezes solid** again. These changes can be made over and over again, as many times as you like. Evaporation and boiling are reversible too. When hot steam is cooled it **condenses** to liquid water.

The photographs on this page illustrate some reversible changes in practice.

The chocolate sauce has a low melting point. It is liquid at room temperature. When it is poured on cold ice cream it sets (freezes) solid. The chocolate melts again in your warm mouth, and tastes delicious!

Liquid crystals are a special state of matter – they are neither a proper solid nor a proper liquid. Some liquid crystals are coloured, and some of these change colour when they are heated. The change is reversible. Different kinds of liquid crystal become brightly coloured at different temperatures. They can be used to make a thermometer display.

A liquid crystal thermometer

The temperature of the ice cream is lower than the melting temperature of the chocolate sauce.

Permanent

Changes of state are reversible, they can be undone. But other kinds of change to materials are permanent, they cannot be reversed. When you toast a slice of bread the change is permanent. Toast does not turn back into fresh bread as it cools down.

The ingredients of a cake change permanently when they are cooked in the oven.

Heat changes soft white bread into crisp brown toast.

Burning is a permanent change too. When candle wax burns it combines with oxygen to make the gas carbon dioxide and water vapour. These gases escape into the atmosphere.

Cement and plaster change permanently when they are mixed with water. They react with the water and set hard.

Baking causes permanent changes.

Burning candles change state permanently – much of the wax disappears.

Once hard, cement will not go soft again.

MAKING CANDLES

Candle wax

Candles were invented in ancient times. The Ancient Egyptians made candles and candle stick holders. Until the 19th century most candles were made from tallow or beeswax. Tallow is hard white animal or vegetable fat. Then in 1859 the first oil well was drilled. Scientists soon discovered how to separate **paraffin wax**, **liquid** paraffin, petrol and other products from the oil. Modern candles are almost always made from paraffin wax.

Ancient Egyptian candleholders

How a candle works

Many people think that it is just the candle wick that burns. But really the main fuel for the candle flame is the wax that surrounds the wick. The heat of the flame melts the wax. The liquid wax soaks up the wick and evaporates. It is the wax vapour that burns with a flame. Some molten wax may drip from the candle and be wasted, but most of the candle wax burns away.

Heat from flame melts wax

Molten wax soaks up wick, evaporates and burns.

Solid wax

How a candle 'works'

A night-light is an efficient candle design. The metal holder stops the molten wax from running away, so it all burns.

Make your own candle

You can purchase a kit or the supplies to make a decorative candle from a craft shop.

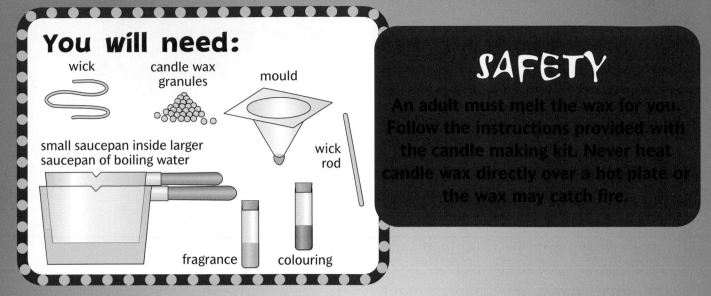

You will need:

wick

candle wax granules

mould

wick rod

small saucepan inside larger saucepan of boiling water

fragrance

colouring

Instructions

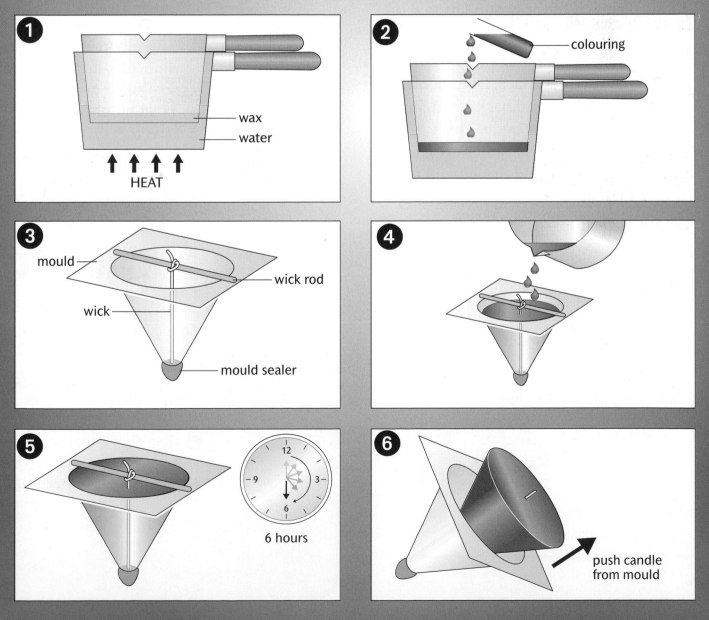

1 wax / water / HEAT

2 colouring

3 mould / wick rod / wick / mould sealer

4

5 6 hours

6 push candle from mould

Water cycle

Water is a very special substance. It is the only substance found on Earth in large amounts as a **solid**, a **liquid** and a **gas**. More than two thirds of the Earth's surface is covered with liquid water in the oceans and seas. Vast quantities of snow and ice cover the two poles and high mountain peaks around the world. The Earth's atmosphere is full of **water vapour**. Sometimes the vapour **condenses** to form clouds then falls to the ground as rain or snow.

The Earth's water is never still. It moves around the Earth's surface, from ocean to atmosphere to land and back, changing state in a never ending cycle. This is the **water cycle**. Without the water cycle to create rain, rivers would dry up and the land would be dusty and barren.

The water cycle is part of the whole process that we call the weather. It is powered by the heat of the Sun. The main features of the water cycle are **summarized** by the diagram opposite. The diagram tells the story of a single water drop.

Where is the world's water?	
oceans	97%
ice caps and glaciers	2%
lakes, rivers, the atmosphere and groundwater	1%

❷ Winds carry the vapour over the land. The wind blows up over the mountains. The air cools as it rises. The drop condenses in a cloud.

❶ A water drop **evaporates** from the ocean. It becomes vapour in the atmosphere.

❻ The river continues to the The drop evaporates in the he Sun and the cycle starts again.

3 The drop falls from the cloud during a rainstorm. It lands in a mountain stream.

4 The drop flows downhill with the stream, passing through rapids and over waterfalls, until it joins a river.

5 The river flows into a lake. At the end of the lake the drop flows over a dam.

Time for action!

Stop global warming!

Global warming will cause more of this:

Flooded homes in India

And this:

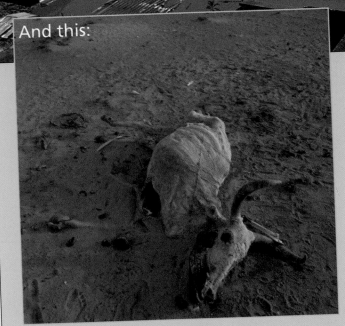

Drought and famine in Africa

You must act to prevent **global warming.**

FACT: The world is getting warmer

- Global warming is caused by human activity

- If we do not act now lives will be lost and communities destroyed

Burning petrol produces carbon dioxide – cut down driving!

What's the cause?

The greenhouse effect causes global warming. Carbon dioxide and other greenhouse gases trap the Sun's heat in the atmosphere. Greenhouse gases released by burning coal, oil and gas in power stations and vehicles are increasing the global greenhouse effect. The more energy we use, the more greenhouse gases we release.

What's the evidence?

The graph (above right) shows that the Earth's average temperature is rising. Research shows that the rising temperatures are linked to increasing human energy consumption.

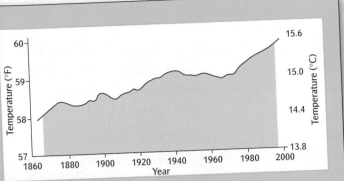

Graph showing how the mean (average) temperature is rising

What are the consequences?

If global warming continues at its present rate droughts and crop failures will become more and more frequent in Africa. Deserts will spread. Millions will starve. Enormous quantities of ice will melt at the poles. Sea levels will rise. Coastal communities around the world will be devastated by floods.

What can you do?

- Conserve energy. Do not leave lights burning or travel by car when you can walk or cycle.
- Use energy efficiently. Make sure your school and home are well insulated. Don't fill the kettle to the brim to make one cup of coffee.
- Find out about alternative energy sources. Solar panels and wind generators do not produce greenhouse gases.
- Ask local politicians to explain what they are doing about global warming. How are they making it easier for local people to conserve energy?

Act now!

Tomorrow it will be too late!

I disagree!

- I'm only one person. Anything I do can't make a difference to global warming.
- It's not what we do in Europe that causes the problem. Americans use far more energy than us.
- In the future people in developing countries will want to use as much energy as Americans and Europeans. What right have we to say they shouldn't?

Glossary

astronomy The study of the planets, stars, galaxies and other objects that make up the Universe

boil A liquid boils when it changes state rapidly to a gas at a constant temperature – its boiling point. Bubbles of a gas grow inside the liquid and rise to the surface

condense To change state from a gas to a liquid. Gases condense when they are cooled to a temperature below the boiling point of the liquid

data-logger A measuring instrument linked to a computer to record readings, such as temperature, over a period of time

dew Droplets of water condensed overnight from the air onto grass, leaves and other cold surfaces

evaporate To change state from a liquid to a gas. Water and other liquids evaporate at all temperatures – the higher the temperature, the faster the rate of evaporation

flammable A flammable material is one that can be ignited by a spark or a flame, and then continues to burn

freeze The change in state from a liquid to a solid. Liquids freeze when they are cooled to a temperature below the melting point of the solid

gas The state of matter in which particles are widely separated and move around freely. The air is a gas

geyser Geysers appear when underground water boils, and the steam expands until it shoots out of the earth in a column

global warming The increase in the average temperature of the Earth year-by-year caused by greenhouse gases that trap heat in the Earth's atmosphere

gravity The force that pulls all objects towards the centre of the Earth, giving them their weight

iceberg A large mass of ice floating in the ocean; icebergs break away from glaciers and the polar ice caps

lava The molten rock that flows from a volcano

liquid The state of matter in which the particles are packed closely together but move freely. Water from the tap is a liquid

liquid crystal An unusual state of matter in which the particles are neither as free to move as in a liquid, nor as fixed in their positions as in a solid. Liquid crystals are used to make thermometer and watch displays

matter We describe all substances and materials as consisting of 'matter'. Matter occupies space and has mass

melt To change state from a solid to a liquid; solids melt when they are heated to their melting point

paraffin wax A greasy white solid extracted from crude oil, used to make candles

porous A porous material is not waterproof; it allows water to soak through

pure ice Ice consisting only of frozen water, containing no dissolved salt or other impurities

solid The state of matter in which the particles are packed closely together in fixed positions

summarize To give a brief account, highlighting the important features of a discussion, a document or some other source of information

volatile A volatile substance is one that evaporates rapidly

volcano A hill or mountain through which lava and hot gases escape from beneath the Earth's crust

volume The amount of space that an object occupies

water cycle The process by which water circulates through different parts of the Earth: from the oceans, to the atmosphere, into rivers and lakes and back again

water vapour Water in the atmosphere in the gas state

Bibliography

Non-fiction

BBC Fact Finders: Water, Chris Ellis
ISBN: 0563346167

Making and Breaking, Science with Solids and Liquids, David Glover
ISBN: 1856970574

Discovery World: Changing Materials, David and Penny Glover
ISBN: 0435339729

Changing from Solids to Liquids to Gases: 5d (Science@School), B.J. Knapp
ISBN: 1862141606

Material Matters: Changing Materials (Shooting Stars), Robert Roland
ISBN: 1841384410

Explore Science: Materials and Their Properties (Explore Science), Angela Royston
ISBN: 0431174458

Material World: Changing Materials (Material World), Robert Snedden
ISBN: 043112101X

Material World: Solids, Liquids and Gases (Material World), Robert Snedden
ISBN: 0431121001

Fiction

'The Legend of Daedalus and Icarus', for example found in:

Tales of the Greek Heroes, Roger Lancelyn Green
ISBN: 0140366830

The Iron Man (Faber Children's Classics), Ted Hughes
ISBN: 0571207618

Internet

www.learn.co.uk/default.asp?WCI=Unit&WCU=8964

www.bbc.co.uk/education/revisewise/science/materials/index.shtml

www.techniquest.org/temperature.htm

www.educate.org.uk/teacher_zone/classroom/science/5d_book/index.htm

http://galileo.imss.firenze.it/museo/4/

Index